The Firefly's Book

By
Brett Ortler

Adventure Publications, Inc.
Cambridge, MN

Dedication

For my son, Oliver, my niece, Charlie, and my wife, Kayli

Acknowledgments

Many thanks to Gerri Slabaugh for making this book possible, to Lora Westberg for the wonderful design, and to the Museum of Science in Boston for allowing me to adapt the data from their wonderful Firefly Watch Project (https://legacy. mos.org/fireflywatch). Thanks are especially in order to Marc Check, my contact person with the Museum, and Don Salvatore, who was kind enough to review this book for accuracy.

I'm also indebted to a number of researchers and scientists for their fine work. This list includes Professor Steve Kay and his colleagues, Professor Helen Ghiradella and her colleagues, Professor Jennifer Frick-Ruppert and her colleagues, Professor Stephen Luk, Stephen Marshall and Marc Branham and Professor Xinhua Fu. Citations of the papers referenced in this book can be found in the Bibliography, which starts on page 95. Several researchers were kind enough to allow me to use their images; for credits, see page 103.

Full citations of the papers referenced in this book can be found in the Bibliography, which starts on page 95.

Photo credits listed on page 103.

Book and cover design by Lora Westberg

Frequently Asked Questions about Fireflies

What are fireflies? Even though fireflies are called fireflies and glowworms, they aren't flies or worms. Fireflies are insects; they belong to a specific group of insects called beetles (see page 10).

How do fireflies light up? Chemistry! Like other bioluminescent animals, fireflies are chemists! They light up because of special chemicals that combine in just the right manner.

Why do fireflies light up? For two reasons: to attract a mate and for self-defense. Fireflies flash to identify themselves to mates. They also flash to let predators know that they contain a toxin and aren't worth eating.

How many firefly species are there? Lots! There are somewhere around 2,000 species worldwide, with about 150 species in North America. Not all are bioluminescent, though; some are active during the day and don't glow at all.

What do fireflies eat? It depends on the age: firefly larvae are predators and they eat all sorts of other critters, even snails. As adult fireflies only live a short time, it's not clear what they eat—if they even eat at all.

Where can I find fireflies? Fireflies are primarily found in the eastern part of North America. Fireflies often prefer dark, wooded areas near water or open, moist fields; this makes state and national parks good places to look. Before you head out on a road trip, though, start by looking in your backyard!

Can I catch fireflies? Yes! Catching fireflies is a great way to learn about them—but you have to be careful, as fireflies are fragile. Only keep them in a jar (see page 76) for a few hours, then release them. Also be sure to put a moist paper towel in the jar with them; otherwise the fireflies might dry out.

Why don't I see as many fireflies as I used to? If you aren't seeing as many fireflies as you used to, you're not alone. Firefly populations appear to be dwindling. While scientists aren't certain what is to blame, it's likely that light pollution and insecticides are playing a role.

What's the best way to identify fireflies? Unless you're a professional, identifying fireflies during the day is tricky. The easiest way to identify firefly species is by their flashing patterns; most species have their own unique flashing patterns, so if you learn to spot it, you'll know which species you are seeing. **Note:** the vast majority of the fireflies photos in this book are not intended for identification purposes; for species identification, pick up a field guide and start exploring!

Table of Contents

Introduction

There are millions of insect species in the world, and it's likely there are many, many more to be discovered. The sheer number of insect species leads to an amazing amount of diversity. From the hated mosquito to the celebrated butterfly, insects truly come in all shapes and sizes. But perhaps no insect species is as famous—or as beloved—as the firefly.

During the day, fireflies are relatively unimpressive, but at night, they literally light up the night, captivating observers of all ages. Chock-full of fascinating facts, photos of fireflies from all over the U.S., and tips about how to catch (and observe) fireflies, this book is intended to be a fun, casual introduction to the world of the firefly. Whether you use it at the cabin, around a campfire at a campground, or in your own backyard, this book is your way to learn more about our summertime six-legged friends.

Memories of Fireflies

I first saw a firefly when I was a kid. We were up north, and it was the middle of summer vacation. The first time I saw a firefly blink, I thought I was seeing things. But then another blinked, and another. I spent the rest of that night running after those tiny blinking lights. Watching fireflies became a summertime tradition, and it still is. Summer just isn't the same without the sky full of fireflies.

The Basics

We almost always refer to them as "fireflies" or "lightning bugs," but fireflies aren't actually bugs or flies. They're beetles.

For those of you needing (a brief!) refresher on your biology: scientists use the taxonomic system to classify and catalog organisms. There are eight categories, or ranks, in it. The general idea is pretty simple: the lower the rank, the more specific you get.

The details aren't that important for our purposes, but beetles —and fireflies—belong to the order Coleoptera. From the venerated scarab to the lowly dung beetle, all beetles belong to the Coleoptera order, and there are over 300,000 in all. (True flies and bugs belong to their own respective orders.[1])

Domain - Eukaryota (Eukaryotes)

Kingdom - Animalia (Animal)

Phylum - Arthropoda (Arthropods)

Class - Insecta (Insects)

Order - Coleoptera

Family

Genus

Species

Insects Are Everywhere

All insects, including fireflies, have three main body parts: a head, a thorax and an abdomen. They also have three sets of legs and one or two pairs of wings. While it's not clear exactly how many insect species there are on Earth, we've cataloged at least one million species, and there may be up to 29 million[1] more. By some accounts, insects represent something like two-thirds of all life on Earth.

Head

Thorax

Abdomen

Fun Fact

Many insect species are discovered each year, and not all of them are found in far-flung locales. In 2013, researchers at the University of Missouri discovered a new insect species—one that lives only on the college campus. Appropriately enough, it is named *Aphis mizzou*[2], in honor of the university.

Fireflies Around the World

There are about 2,000 species of fireflies in the world, and they are found on every continent except Antarctica.

About 150 species of fireflies are found throughout North America, but despite their name, not all fireflies glow. Sadly, in the U.S., only scattered pockets of luminescent fireflies are found west of the Rocky Mountains, though many non-glowing species are found there. Scientists aren't exactly sure why.

Thankfully, the rest of the country has dozens of varieties of glowing fireflies for us to enjoy.

Antarctic midge

Fun Fact

Remember how there are millions of insect species in the world and they make up most life on Earth? Poor Antarctica has only one insect species, the Antarctic midge, which is the only terrestrial animal on the continent. (It also happens to be pretty ugly.)

Firefly or Glowworm?

Fireflies are often referred to as glowworms, and at first glance, fireflies (and especially their larvae) do look similar to small earthworms or inchworms. Nonetheless, while there *are* a few bioluminescent worm species, worms are very, very different from insects. They belong to an entirely differently phylum (see page 10), which means that their body structure is completely different. Even the critters we often call "inchworms" aren't actually worms; they are caterpillars.

An "inchworm," which is really a moth larva

An earthworm

Fun Fact

The firefly-or-glowworm problem is exactly why scientists refer to particular organisms by scientific names. This system, called binomial nomenclature, gives each life-form a unique name that is the same worldwide, helping eliminate confusion.

Fireflies That Don't Glow

A non-glowing firefly

It's a sad truth, but not all adult fireflies glow. Instead of lighting up, these fireflies attract a mate by using pheromones (chemicals intended to attract a mate). Not surprisingly, some of these non-glowing species are primarily active during the day— not at night.

Nonetheless, all firefly larvae have light organs and can glow, and this is true even for species that don't glow as adults. The species *Ellychnia corrusca* is a good example of this. As a larva, it glows, but adults lack lantern organs[1] altogether.

Fun Fact

While pheromones might not seem as flashy as the firefly's flashing patterns, insect pheromones are just as interesting and can resemble something out of a spy movie. The larvae of one species of blister beetle conspire together to mimic the shape (and smell) of a female ground bee. When the male bee arrives, they clamp onto his underside, hitching a ride until he mates with a female. When that happens, some jump ship; others wait to find another female. Eventually, the females bring them back[2] to the nest, which they take over.

What's in a Name?

In North America, there are three primary groups of fireflies: the *Photinus*, *Photuris*, and *Pyractomena* fireflies. (For those interested in taxonomy, each group is its own genus.) Many other firefly genera (the plural of genus) exist in North America, but they are little studied, and not many of those species light up, so most attention is focused on the "big three."

It's often useful to translate the technical terms that scientists use, as the name sometimes tells you something about the group of critters itself. Not surprisingly, scientists named each genus for the firefly's most noticeable characteristic—the fact that it glows in the dark.

Photinus (foe-TINE-us) stems[1] from Greek and means "**shining.**"

Photuris (foe-TOUR-us) also originates[2] from Greek and means "**luminous tail.**"

Pyractomena (py-RAC-toh-ME-na) comes from[3] the Greek word for "**fire.**"

Fun Fact

If you think you've heard some of these words before, you have. *Photinus* and *Photuris* share the same root word as photon and photo, and *Pyractomena* is related to the words "pyromaniac" and "pyre."

The Firefly's Life Cycle

To understand the basics of a
firefly's life cycle, it's best to
start with the familiar sight of
fireflies lighting up a summer's night.
For a firefly, glowing in the dark is beneficial
for many reasons, but it's especially
important for one thing: locating a mate.

Just as bird species have specific mating calls, most
firefly species have their own flashing patterns. Female
fireflies don't fly very often, so male fireflies do the bulk of the
signaling from the air. The female usually situates herself on
the ground or on vegetation of some sort and blinks back
when she spots a male in the air. As each species of firefly is
on the lookout for one specific type of signal, they can spot a
potential mate, even amid a sky full of different firefly species.

Fun Fact
One could argue that humans have copied the firefly's on-again
off-again communication strategy; firefly-like signal lamps have been
used aboard naval vessels for over a hundred years.

Four Phases of Life

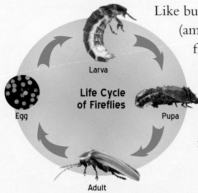

Life Cycle of Fireflies

Egg

Larva

Pupa

Adult

Like butterflies, true flies, and ants (among many other insects), fireflies go through four general stages of life: they start life as an **egg**, which hatches into a **larva**. After feeding, the larva **pupates**, where it transforms into an **adult**.

Fun Fact

It might seem strange, but going through four stages of life gives many insects a number of unique advantages. For example, a monarch butterfly caterpillar has an entirely different diet than a full-grown butterfly.[1] Monarch caterpillars only eat milkweed plants; like many other butterfly species, adult monarchs eat nectar from a variety of different flowers. (This is why people often plant butterfly gardens.)

Firefly Eggs and Larvae

After mating with a male, the female firefly lays[1] her eggs amid moist vegetation, such as moss. Egg production varies by species, but females often produce over a hundred eggs, sometimes several times that. About a month after the female lays her eggs, they hatch, and the larvae emerge and start eating.

Glowing firefly eggs from Asia

A firefly larva

Larvae Lunch

Firefly larvae have one job: eating, and it's something they excel at. Firefly larvae are predators, preying upon a variety of creatures, including other insects, snails and slugs. According to the Ohio State Extension Service, fireflies "inject strong paralytic and digestive juices into their prey and then suck[1] the dissolved body contents." Yuck!

A European firefly
larva feeding on a snail

Why Larvae Glow

Adult fireflies glow for a number of reasons (see page 33), but it's not exactly clear why larvae glow. The larvae of various species glow under different circumstances, but researchers are confident that[1] some instances of larvae glowing occur in self-defense. While lighting up obviously attracts attention, it seemingly scares off enough predators to be worthwhile.

European firefly larva glowing

Fun Fact

According to a researcher[2] at the University of Florida, the larvae of some firefly species are pretty jumpy: some light up after gunfire, while others light up when a football is thrown in their vicinity. Others are even more skittish, lighting up when a door is opened or closed in the room where they are kept.

The Pupal Stage

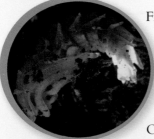

Firefly larvae hibernate underground or on (or under) tree bark. When pupating, some species build a small mud structure around themselves and change into a pupa, where they transform from a larva into an adult. Other species pupate on trees.

From the recent cult hit television show *Firefly* to Keats and the Chinese *Book of Odes*, fireflies have been referenced in art and literature for centuries.

Adult Fireflies Emerge, with a Deadline

Just as firefly larvae are focused on eating, adult fireflies have one goal once they emerge: mating. And they are on a (literal!) deadline, as many fireflies only live a matter of days to weeks, with some species only lasting two or three days as adults.

Home Sweet Home

With so many different firefly species in the United States,
it's no surprise that firefly habitat varies significantly. Some
species are found in wooded areas, while others are found in
more open locations, but if you're looking for fireflies, here's
a tip: look for water. Almost all firefly species tend to live
near some sort of water source. Whether it's a babbling forest
stream or simply[1] a flooded ditch, fireflies tend to prefer
humidity and moist areas. If you live in the right area, you
might even find fireflies in your backyard!

Fun Fact

Speaking of water, some species of fireflies found in Asia are aquatic
and their larvae even have gills! These larvae[2] hunt snails underwater.

Lantern Organs

Fireflies glow because of specialized[1] light-producing organs located on their undersides that are often referred to as lantern organs. Usually, we think of chemical reactions occurring in a laboratory, but fireflies light up because of a special chemical reaction that takes place within their bodies. A firefly's body is a living science lab.

If the firefly's body is the "laboratory" where the reaction takes place, then the lantern organs are the equipment needed for the reaction. The lantern organs are made up of three basic parts: photocytes, peroxisomes and the trachea.

While those terms may sound confusing, it's not hard to understand the basics. The trachea helps trigger the reaction, and it is surrounded by photocytes, which are the "reaction chambers" where the chemical reaction takes place. Peroxisomes are storage tanks that hold the chemicals that help a firefly light up.

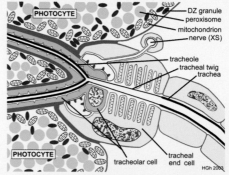

A diagram of a firefly's lantern organs

Luciferin: A Chemical Named for the Morning Star

Fireflies light up in large part to a chemical compound called luciferin. Light is produced when luciferin is combined with luciferase[1] (an enzyme), ATP (a molecule that transfers energy) and oxygen. Luciferin and luciferase get their names from the name Lucifer, which means "light-bearer." Before Lucifer became synonymous with the prince of darkness, the word referred to the planet Venus, which shines so brightly that it's often mistaken as a star—or even a low-flying airplane.

Fun Fact

The first relatively large amount of firefly luciferin was obtained in a lab in 1957. Unfortunately, to get it, researchers had to extract[2] 15,000 firefly lanterns. The end result? A whopping 9 milligrams of luciferin. That's not much: a paper clip weighs 1 gram, about 111 times more.

The Science of Fireflies

While fireflies and other bioluminescent organisms have been admired for centuries, scientific understanding of fireflies (and bioluminescent organisms generally) lagged far[1] behind. People have always noticed them, though. The ancient Greeks cataloged bioluminescent organisms, and later, explorers such as Christopher Columbus and Sir Francis Drake noted the presence of bioluminescent creatures, including some firefly species.

The first scientific progress, however, was made much later, in the mid-to-late nineteenth century, when luciferine and luciferase were first isolated. The golden age of firefly research, however, occurred in the 1950s and 1960s, when many important papers were published, and the essential details of firefly chemistry were published for the first time.

How Do Fireflies Control When They Light Up?

The basic chemical reaction that causes a firefly to light up is now well understood, but for as long as they have been studied, one question has continued to perplex researchers: how do fireflies begin the reaction and control it? After all, many firefly species depend on a precise blinking pattern (sometimes down to the millisecond), and without it, it's likely that they'd experience less success in finding a mate.

The issue still isn't definitively resolved. Nitric oxide[1] and hydrogen peroxide[2] have both been suggested as possible flash "triggers." While the question remains unanswered, we're certainly getting closer to knowing what causes fireflies to "flip their switch" and light up.

Fun Fact

Firefly luciferin synthesis first occurred in 1961, the same year of the first spaceflights of Yuri Gagarin and Alan Shepard, Jr.

How Bright are Fireflies?

Candlepower is a common—albeit old-fashioned—measurement of brightness. As you might expect, one candle produces one candlepower. On a dark summer's night, you might think that fireflies flash as brightly as a candle—or even brighter. But in reality, fireflies are only a fraction as bright. Brightness varies by species, but most fireflies are about $\frac{1}{400}$ as bright[1] as a candle.

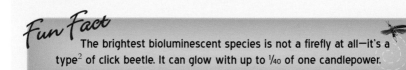

Fun Fact

The brightest bioluminescent species is not a firefly at all—it's a type[2] of click beetle. It can glow with up to $\frac{1}{40}$ of one candlepower.

Why Do Fireflies Look So Bright?

It may be hard to believe that fireflies are only a few hundredths as bright as a candle. After all, they look very bright, and they often seem brighter than a candle. As it happens, the human eye[1] is most sensitive to yellow-green light—and this is what most fireflies emit. We literally can't help but notice them!

Visible light spectrum

The Science Behind the Firefly Color Palette

When you see a firefly glowing, you might disagree with your friends and family about what color it is. Well, researchers at Johns Hopkins and the National Institutes of Health set out to precisely determine the range of light that fireflies emit. As it turns out, firefly light can vary quite a bit.

The science behind this is relatively simple. Visible light makes up a relatively small portion of the electromagnetic spectrum, and it ranges from red on one end to violet on the other. All of the other colors fall in between; each color has a specific frequency range that corresponds to the intensity and the wavelength of light. For example, red light has longer wavelengths and is less energetic; violet light has shorter wavelengths and is more energetic. A handy way to remember the order of the visible light spectrum is to remember the made-up name "Roy G. Biv," which stands for red, orange, yellow, green, blue, indigo, violet.

So how does firefly light measure[1] up? Well, as it turns out, the firefly emissions vary considerably by species, with each species reaching distinct peaks in the electromagnetic spectrum. Most, however, fall somewhere in the green to yellow range.

Mating: Why They Glow

While it's not exactly clear how
fireflies regulate their flashing
patterns, scientists have a pretty
good idea why fireflies light[1] up.
Flashing patterns play a central role
in mating. Males advertise their presence

to females by lighting up, and if the female is interested, she
returns the flashing pattern to signal the male to come closer.

Sometimes It's Good to Be Noticed

Many other creatures depend on camouflage and staying hidden to survive. Not the firefly. Many fireflies have a secret weapon: they are toxic, and they want other species to know about it. As many creatures rely on camouflage to survive, this may seem counterintuitive. This tendency is called aposematism—and it's not restricted to fireflies. Well-known examples include skunks, poison dart frogs, and the ever-popular monarch butterfly.

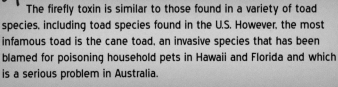

A cane toad

Fun Fact

The firefly toxin is similar to those found in a variety of toad species, including toad species found in the U.S. However, the most infamous toad is the cane toad, an invasive species that has been blamed for poisoning household pets in Hawaii and Florida and which is a serious problem in Australia.

A World of Variation

When people think of fireflies, they often tend to think of a simple on-again, off-again flashing pattern, but as anyone who's watched fireflies knows, firefly flashing patterns vary significantly by species. And those variations can be surprisingly complex: Some fireflies blink like Christmas lights, while others space out their flashes, or flash for longer[1] periods.

Still others flash only after flying in a particular pattern. The Common Eastern Firefly is an example of this. Often called the Big Dipper Firefly, it lights up only after tracing[2] a "Big Dipper-like" path. According to entomologist[3] Marc Branham, in some species, females are attracted to males that blink often—and more brightly than the competition.

The flashing pattern of the Common Eastern Firefly

Fun Fact

Other than actually capturing live specimens (and dissecting them), deciphering a firefly's flash pattern is the primary way to determine which species you're seeing.

Other Creatures with Bioluminescence

Fireflies may be among the most familiar bioluminescent creatures, but they are hardly alone. In fact, many species exhibit bioluminescence, with new species still being discovered. The vast majority are marine[1] organisms, and they range from small organisms, such as plankton and bacteria, to larger creatures, such as squids and even sharks. Bioluminescence in land animals is much rarer, but not unheard of; in addition to fireflies, there are bioluminescent bacteria, mushrooms, and even[2] a bioluminescent snail.

New Zealand is especially famous for its "glowworms," which are actually the bioluminescent larvae of a type of fly (*Arachnocampa luminosa*). These larvae are found in caves, where they hang from the ceiling, creating an otherworldly —and very popular—light show.

Bioluminescent fungi, jellyfish, and New Zealand's "glowworms"

Lunch—and Something Extra

The *Photinus* firefly species contain defensive steroids called lucibufagins, which cause them to taste quite bad (and can even poison creatures that eat them). Fireflies of the *Photuris* genus, however, don't possess these toxins, but they sometimes end up with them anyway. How? Well, they mimic the flashing patterns of *Photinus* fireflies in order to lure the males to their doom. These Femme Fatales get more than just an easy lunch out of the deal, too. According to a study[1] led by the famous entomologist Thomas Eisner, these fireflies actually absorb the lucibufagins, which then protect them!

Fun Fact

Thomas Eisner is known as the "father of chemical entomology" and over his long career, he studied the chemical defense mechanisms of a wide range of "creepy crawlies," including ants, millipedes, beetles, and whip scorpions. His wide-ranging work revolutionized entomology, and it also led to the realization that chemical warfare is quite common in nature. The titles alone[2] of his scientific papers make this clear. Examples include: "Survival by acid defense," "Ploy and counterploy in predator-prey interaction," and "Defensive use of a 'fecal shield' by a beetle larva." He even once wrote a paper about catnip. (As it turns out, catnip helps repel predatory insects!)

Predator and Prey

While we appreciate fireflies for their beauty, other creatures like to eat them for a snack. According to a study[1] by researchers in the Great Smoky Mountains, a number of predators specifically prey upon fireflies. This murderers' row

includes orb-weaving spiders, harvestmen (daddy longlegs), assassin bugs and hanging flies; these insects are seemingly unaffected by the toxins that the fireflies carry.

A male orb
weaver spider

Fun Fact

Fireflies aren't the only creatures that are consumed despite a potent chemical defense. In fact, for every species with such a survival strategy, there is usually another that can circumvent it. Skunks are perhaps the perfect example: their smelly odor deters most predators, but not all of them. Great horned owls regularly feed on skunks, even if the skunks spray them. How do they stand the smell? Easy: they don't have much of a sense of smell to begin with.

Reflex Bleeding: A Strange Survival Strategy

When fireflies are attacked, they have several additional methods of defending themselves. One strategy[1] is referred to as "reflex bleeding," and it's just what it sounds like: when a firefly is in danger, it often begins to bleed—often even if it hasn't been punctured. This might sound like a counter-intuitive defense strategy, but fireflies contain toxins in their blood, and this strategy—which is harmless to the firefly—advertises just how distasteful they are. It's the equivalent of giving a dinner guest a terrible appetizer so they forget about the main course!

This reflex bleeding is useful for another reason, too; firefly blood doesn't just taste bad, it's also sticky and can "glue" an attacker's mouth shut.

Fun Fact

Scientifically speaking, the strategy of playing dead[2] is referred to as thanatosis, which is named for Thanatos, the Greek god who personified death.

Don't Feed Fireflies to Your Pet...or Your Brother!

Even though they contain a toxin, handling fireflies (putting them in a jar, say) is entirely safe. But don't let your pet—or a little one—eat them on accident. First of all, doing so would undoubtedly be unpleasant, as they taste very bitter. Worse yet, they are toxic. According to a report[1] from Cornell University, there are two confirmed cases of pets dying after ingesting fireflies. The deaths occurred in two bearded dragon lizards (*Pogona* genus) and occurred after ingesting only a handful of fireflies. While it's unlikely your critter would share the same fate, better safe than sorry!

Fun Fact

Fireflies aren't the only organisms that survive by tasting bad. Many plants have chemical defenses to prevent them from being eaten. Oak trees are one example; their acorns contain tannins, chemicals that make acorns taste very bitter. Even this doesn't ensure that acorns won't end up as lunch; squirrels certainly don't seem to mind the tannins. (Then again, it's not all bad for the oak trees. Squirrels often bury acorns in preparation for winter, and they don't always find the ones they bury, acting as inadvertent tree-planters.)

A Cold Light

Chemical reactions that emit energy
(often in the form of heat) are
called exothermic reactions.
Examples include the heat
and light from a campfire or the heat
and energy produced in an automotive
engine. When fireflies produce light, they
are almost perfectly efficient, producing
very little heat. For this reason, they are
often said to emit a "cold light." As it turns[1]
out, that's a good thing. If their chemical
reactions were less efficient and produced
more heat, it's likely they'd overheat and die.

Memories of Fireflies

Like Fourth of July fireworks, cool swims on hot days, and lazy vacations, fireflies are a sign of summer. Many of us have cherished memories of capturing fireflies and placing them in Mason jars, creating dim, but beautiful, "night lights" to enjoy. While firefly light might be what makes them most famous, there is even more to fireflies than that. A firefly's life is a struggle for survival defined by danger, romance, and even deception.

Common Fireflies

Unless you're trained as an entomologist, identifying an individual firefly species is difficult. In fact, even field guides usually only list a few individual species. That's because there are simply too many species to include, and it's too difficult to differentiate among them. (Doing so often requires dissecting the fireflies!)

Thankfully, this book isn't about identifying firefly species —it's about enjoying them, so we'll only cover the basics. Generally speaking, there are three primary types of bioluminescent fireflies in the United States: the *Photinus* fireflies, the *Photuris* fireflies, and the *Pyractomena* fireflies. Each belongs to its own genus, and rightfully so, as each type is unique.

The Photinus Fireflies

If you see a firefly, there's a good chance it belongs to the *Photinus* genus, which includes dozens and dozens of species. In fact, the most common firefly species in the United States—the Common Eastern Firefly—is part of this group. Famous for its characteristic flight pattern, the Common Eastern Firefly is also known as the Big Dipper Firefly, because it flies in a J-shaped pattern that resembles the famous nighttime pattern in the sky.

About a half inch in size, *Photinus* fireflies are found throughout the eastern portion of the United States; they glow a yellow-green color.

A *Photinus* firefly

The Femme Fatales

The *Photuris* genus includes about 20 species, but this group is far more famous for its common name—the Femme Fatales. This name is apt, as the

A *Photuris* firefly.

females of this species mimic the flashing patterns and colors of *Photinus* females in order to draw *Photinus* male fireflies closer. When the *Photinus* males approach, the *Photuris* female attacks, devouring the unsuspecting male. (Talk about a rough day: instead of mating, the males get eaten!)

This type of deception is called aggressive mimicry.

At about an inch[1] long, *Photuris* fireflies are larger than the other two types of fireflies, and they also glow a darker shade of green.

Fun Fact

The *Photuris* fireflies are not the only species that are aggressive mimics; the anglerfish is perhaps the most famous aggressive mimic. In the absolute darkness of the deep ocean, it dangles a lighted "lure" that leads right to its mouth—and its large teeth. It was even featured in *Finding Nemo*!

The Femme Fatales Find a Mate

The Femme Fatales aren't the only mimics in the night, however. To signal a mate, the *Photuris* males actually mimic the flashing pattern of the *Photinus* fireflies. So not only do the Photuris females mimic the "other guy"—the males do, too! They do so because this tricks the *Photuris* female into signaling and revealing herself. So when a male approaches[1] the female, it's not immediately clear whether she's getting dinner or a date.

The *Pyractomena* Fireflies

Roughly the same size as *Photinus* fireflies, *Pyractomena* fireflies are notable because they glow in shades of yellow or orange and are often said to resemble[1] the color of an ember from a campfire. There are about 18 species in this genus and they are found throughout North America.

A *Pyractomena* firefly.

A Synchronized Show in the Great Smokies

While fireflies can be found over much of the eastern half of North America, some places are firefly destinations and draw tourists from around the world. Great Smoky Mountains National Park is the most famous example in the United States, and it draws thousands of visitors in early May and June every year. The park doesn't just have fireflies: it features a species of synchronous[1] fireflies in which all the fireflies in the group light up at once.

The event is so popular that the park's website even lists suggestions for firefly etiquette. Flashlights and car headlights are frowned upon, as they interfere with the fireflies and ruin the spectators' night vision; instead, the park suggests that visitors bring flashlights covered in blue or red cellophane. This doesn't bother the fireflies and preserves the viewers' night vision.

Why Synchronize?

In a forest full of synchronized fireflies, one might[1] think that it might be hard for a female to spot a mate. After all, if all of the other males light up at the same time, it seems easy to lose them in "the crowd."

But the reality may be just the opposite. From a female firefly's point of view, when fireflies don't synchronize, a dark evening can be something of a cluttered mess, with males of many firefly species flashing at once. This can make it difficult for her to find a male of her own species. Synchronized fireflies don't have this problem.

Fun Fact

There are other such firefly destinations around the world. A number of Japanese[2] cities hold *hotaru matsuri*—firefly festivals—each year: at some of these festivals, captured fireflies are released just for the occasion.

Other Synchronous Firefly Varieties

While they make Great Smoky Mountains National Park a tourist destination, synchronous fireflies are far more common in Southeast Asia[1], which is home to several species in the genus *Pteroptyx*. Unfortunately, these species have been little studied and are now at risk because of habitat destruction.

The Blue Ghost Firefly: A Steady Light in the Dark

Most luminescent firefly species in North America have a specific flashing pattern. This "calling card" enables males to advertise their presence to females of the same species; synchronized fireflies have a similar strategy, except all males of the species light up at the same time.

The Blue Ghost Firefly is notably different. It belongs to an entirely different genus (*Phausis*), and it doesn't blink at all; instead, it emits a continuous glow. This glow can vary from a few seconds up to a minute[1] or more. Compared to other fireflies, comparatively little is known about this species (and even its genus).

The Blue Ghost Firefly

Fun Fact

The Blue Ghost's name is a bit of a misnomer: its light looks blue from a distance but is bright green[2] close up. The "ghost" part of the name is accurate, though, as Blue Ghosts are positively tiny—about the size of a grain of rice—and they blaze a wispy, wraith-like trail of light through the night.

Concerns for the Present (and the Future)

To thrive, fireflies often require very specific habitat—usually undisturbed moist, wooded areas or open, moist fields—but such land is becoming rare due to human development. Not surprisingly, so are fireflies. It seems to be a worldwide problem and has been noticed everywhere from the U.S. to Thailand.[1] Fireflies have even outright disappeared in some areas.

Along with habitat loss, scientists think that light pollution is also part of the problem. Given the universal presence of electric lights, the night sky is often very brightly lit: in major urban areas, only a handful of stars are visible, and one can easily read by the background light. Fireflies simply can't compete against such a well-lit background, and as lighting up is central to their reproductive strategy, it's not surprising that fireflies might have trouble reproducing.

Worse yet, urban areas often have more light pollution and more land development. For fireflies, these two problems are a double whammy: they have fewer places to live and more glare.

Problems with Insecticides

The decline in firefly populations may have another factor: insecticides. Broad-spectrum insecticides are available at any hardware or home-improvement store, and overuse of these products harms many other insects. By definition[1], broad-spectrum products affect many species in addition to the target species. Fireflies are almost certainly affected as well.

While insecticides have obvious benefits in agricultural settings, they are often used more or less cosmetically on lawns and gardens. According to a report[2] by the Xerxes Society for Invertebrate Conservation, it is exactly this cosmetic use that is most damaging, as the amount of product used is generally higher, enough to kill bees outright.

Fun Fact

Thankfully, in a backyard setting, many insect and pest problems can be dealt with using other products/strategies. For tips, visit[3] the website for Purdue's Pesticide Program: www.ppp.purdue.edu/Pubs/ppp34.html

A Lot We Still Don't Know

Research done using luciferase

Given how popular fireflies are, one might think that we'd know everything there is to know about them. But that's actually not the case; we still have a lot to learn. Efforts to describe even some of the most interesting species—including the Blue Ghost —have begun only relatively recently.

For example, while we know how fireflies light up, it's not definitively settled how they control their flashing patterns so precisely. While that might not seem like an important research question, it's important to remember that an advance in one area of science can help other parts of science progress. (That's part of the fun part of science: you don't know what one discovery could lead to next.)

So far, firefly research has had a number of useful applications. Scientists adapted firefly research to help create brighter LEDs (light-emitting diodes). In medicine, researchers have "painted" genes with luciferin, the chemical that makes fireflies glow. When the glowing genes were inherited, they still glowed, allowing researchers to quickly determine which genes were inherited. Similarly, biologists used luciferase[1] to understand plant growth.

Mythology

Just as fireflies are perennially popular with scientists and laypeople alike, they've been a popular subject in myth and legend across many cultures. Japan[1] has arguably the closest connection to fireflies. There, fireflies are referred to as *hotaru*, and they  were once said to represent the souls of dead. Alternately, fireflies are also seen as a symbol of love. Japanese art history is also replete with paintings of fireflies.

The country's close connection with fireflies continues; today they are viewed as a symbol of the environmental movement.

Fun Fact

Not all cultures viewed fireflies positively. According to[2] Purdue Extension, it was once thought that if you got a firefly's "fire" in your eye, you'd go blind in that eye. On a similar note, an old European legend attested that if a firefly entered the house, someone in the house was bound to die.

Fireflies, Poets and Playwrights

Given their fleeting beauty, the firefly is a perfect fit for the poet. It's a tailor-made simile. Perhaps that's why references to glowworms and fireflies are common in fine literature, with references by the likes of William Blake, John Keats and William Wordsworth. William Shakespeare himself makes three distinct references to "glow-worms," the best of which arguably is:

> Where now his son's like a glow-worm in the night,
> The which hath fire in darkness, none in light[1]
>
> *Pericles, Prince of Tyre*

More Popular Than Ever

Even though firefly populations seem to be dwindling, fireflies are more popular than ever. References to them in popular culture aren't hard to come by—a famous science fiction franchise (*Firefly*) is named after them, the song "Fireflies" topped the charts in 2010, and they are a staple of children's toys and children's literature. From Eric Carle's classic *The Very Lonely Firefly* to firefly-themed night lights, fireflies are likely to remain a popular cultural symbol.

But what does this mean for firefly populations? Admiration alone doesn't ensure survival. (Just ask other beloved, but endangered, animals.) While the situation is nowhere near as dire as it is with the mountain gorilla or the giant panda, firefly populations certainly seem to be in trouble. Stabilizing the population will likely require harnessing our admiration for fireflies.

Fun Fact

The science fiction show *Firefly* gets its name from its ship's resemblance to the firefly!

Memories of Fireflies

For just a moment, consider just how many children grew up with wonderful memories of chasing fireflies on warm summer evenings. Then think of how many adults continue to enjoy fireflies each summer, and it's easy to understand why fireflies are such a joy to so many people.

Watching Fireflies

Watching (and catching) fireflies is a time-honored summertime tradition, but it's not always easy to know where or when to look. Fireflies are generally found in specific habitats, and firefly season only lasts a short time—often a matter of days or weeks—so it's not only important to know where to look, but when. This section of the book covers the basics of what you need to know, including firefly distribution in the United States, information on recognizing prime firefly habitat, a calendar for determining firefly "season" where you live, and a detailed chart of firefly patterns to help you attempt to identify the fireflies you spot.

Firefly Distribution

Fireflies are generally found throughout the eastern half of the United States, and there are ample opportunities to observe them, with dozens of species to enjoy and identify. The map shows the rough distribution of fireflies. Note: fireflies do live in the western half of the continent, but these species generally don't light up, instead attracting mates via[1] pheromones. Scattered pockets of luminescent fireflies can be found on occasion, however.

The above map[2] was produced with data collected regarding firefly sightings from the Firefly Watch Project via Boston's Museum of Science (https:// legacy.mos.org/fireflywatch/). Note that this is a generalized map only and is not intended to be a guide to finding specific species; the map is also not comprehensive.

Season Calendar

One of the primary reasons fireflies light up is to attract a mate, so it stands to reason that they only light up during their mating season. (While it's true that larvae also light up, they are much, much more difficult to find.) Thankfully, not every firefly species shares the same mating season; this means that the various species in a given habitat overlap, making it possible to observe fireflies over the course of several weeks, and in some areas, over several months. This range depends a great deal on your location's climate and habitat. In the southern portions of the country, which tend to be warmer, fireflies are visible during more of the year.

This calendar, derived from data collected by the Firefly Watch Project[1] at Boston's Museum of Science (https://legacy.mos.org/fireflywatch), consists of firefly-sighting data provided by volunteers from across the country. It lists the first and last dates fireflies were spotted. As it consists of volunteer data, this calendar is necessarily incomplete and doesn't cover every state/province. While certainly not comprehensive, it should give you a (very) rough idea of when to look for fireflies in your area.

May 11
Sep 2

Data Not
Available

Jun 8
Jul 21

Jun 12
Aug 15

Jun 15
Aug 21

Jun 3
Aug 2

Apr 14
Sep 5

Data Not
Available

May 19
Aug 25

May 23
Sep 9

May 26
Sep 30

Apr 30
Sep 6

May 20
Aug 24

Jun 30
Aug 9

May 6
Sep 23

May 1
Aug 21

Apr 30
Sep 25

May 15
Oct 9

Jun 1
Sep 4

May 1
Aug 11

Apr 19
Oct 5

Mar 25
Sep 6

Apr 29
Sep 19

Apr 15
Oct 5

May 17
Sep 8

Apr 28
Sep 11

Apr 20
Sep 11

May 5
Oct 9

Apr 8
Oct 24

Jun 10
Jul 9

May 12
Jul 9

Mar 15
Oct 1

May 13
Sep 28

May 15
Sep 11

Mar 26
Sep 28

Mar 3
Sep 26

Mar 12
Sep 30

May 18
Aug 31

Mar 20
Sep 27

Mar 11
Sep 15

Feb 13
Oct 10

Apr 11
Oct 23

Mar 8
Nov 25

Finding the Right Habitat

Once you know where to look geographically and roughly when to start looking, you still need to find firefly habitat. As there are many different firefly species, they are found in a variety of habitats, but there are a few general tips to help you start searching:

- Many firefly species are often found near some variety of a water source, whether it's a pond, a depression, a stream, or a lake.

- Fireflies are often found in open areas[1] near woods, such as meadows or clearings.[2]

- You don't necessarily need to leave home to find them. If you live in an area with a water source and some woods, take a look around!

- If your neighborhood doesn't have them, check out a local, state or national park near you. Some parks even have specific firefly-watching programs!

- Areas that haven't been sprayed with insecticides or treated with fertilizers are more likely to foster fireflies.

- Dark areas are best, so be sure to turn off that flood lamp on the porch!

- If you want help finding fireflies, consider contacting the experts! Local science museums and university extension services are often great resources of information, and they often sponsor fun hands-on activities!

Firefly Flashing Patterns

A firefly's flashing pattern isn't just a pretty pattern—if you pay really close attention, a firefly's light show can tell you a lot about the firefly, including its species and even how warm or cold it is outside.

Because firefly flashing patterns are complex, it's best to start with the basics. The following three charts[1] depict the flashing patterns of a few common firefly species, and each chart shows one special aspect of firefly flashing patterns. The first shows how much flashing patterns vary by species. The second chart shows how different male and female flashing patterns are from each other. The third shows how temperature affects how often fireflies light up. Once you're familiar with these three concepts, you're on your way to identifying fireflies by their

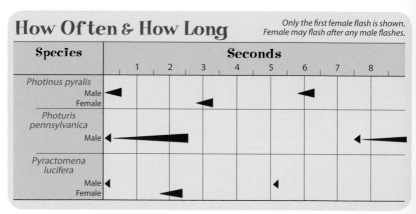

For a more complete chart of flashing patterns, visit: https://legacy.mos.org/fireflywatch/flash_chart

blinking patterns alone. It's certainly not foolproof, but it's
definitely a fun way to spend a summer's night.

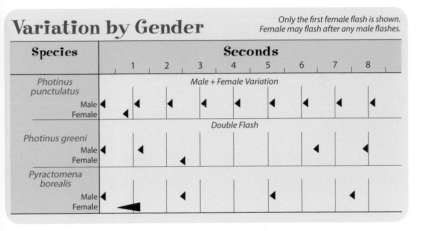

Variation by Gender

Only the first female flash is shown.
Female may flash after any male flashes.

Species	Seconds							
	1	2	3	4	5	6	7	8
Photinus punctulatus			*Male + Female Variation*					
Male	◄ ◄	◄	◄	◄	◄	◄	◄	◄
Female	◄							
Photinus greeni			*Double Flash*					
Male	◄	◄					◄	◄
Female			◄					
Pyractomena borealis								
Male	◄		◄		◄		◄	
Female		◄						

Variation by Temperature

Only the first female flash is shown.
Female may flash after any male flashes.

Species	Seconds							
	1	2	3	4	5	6	7	8
Photinus punctulatus								
male 73°	◄					◄		
male 71°	◄						◄	
male 67°	◄						◄	

All charts based on data from *Studies on the Flash Communication System in Photinus Fireflies*, James E. Lloyd, University of Michigan Museum of Zoology, 1966.

71

Memories of Fireflies

Usually, we view nature from a distance. This is often a good thing, as you wouldn't want to get too close to the business end of an eagle's talons or a buck's antlers. But there is something to be said for watching nature up close. Fun, fascinating and beautiful, fireflies are perhaps the perfect animals for the budding scientist (young or old!) to observe. In this respect, something as simple as a Mason jar full of fireflies can open the door to a lifetime of wonder.

Fun Firefly Projects

Of course, part of the fun of watching fireflies is that you can catch them and view them up close! As long as you're careful and don't harm the fireflies, this is a great way for kids and adults alike to have fun, learn about science and enjoy a summer's night. Thanks to the existence of Citizen Science Projects, your observations might even help advance our scientific understanding of fireflies and just might help the fireflies themselves.

Project 1

A Home Away From Home

> **Equipment[1]:**
> - Clear glass jar
> - Moist paper towel
> - Grass and twigs

Steps:

1. Use a clear glass jar, and preferably a tall one, which will give your firefly room to roam around.

2. Put some grass and twigs in the jar, along with a moist paper towel.

3. Be sure to let your fireflies go! Adults of some firefly species only live a matter of days, and given the apparent decline in firefly populations, it's important to get every firefly bachelor out there.

Project 2
Do-It-Yourself Bioluminescence

Sadly, firefly season doesn't last all year long, so what do you do when there aren't any fireflies to watch? Why not culture some bioluminescent bacteria? Kits with bioluminescent bacteria are available for relatively cheap prices online (around $25) and make for exciting science fair projects and class activities. The organisms don't usually stay lit up for more than a day or two, but they're a great way to introduce kids to the basics of bioluminescence.

Plus, it's also plain-old fun. Manufacturers of these kits vary, but check with the major science education providers, such as Carolina (www.carolina.com), Sea Farms (http://seafarms.com) and EMP Co. (http://empco.org/).

TIP: If you're really inventive, consider using your glowing experiment in a (short-lived) craft, such as a glowing night light.

Project 3
Firefly Photography

When you've got your fireflies in a jar, it's a perfect time to take some photographs of them! Try taking some long exposure shots of them (several seconds, say). If there are many fireflies in the vicinity, try taking some longer exposure shots of the entire area. As the fireflies light up and then blink out, they leave bending and warping trails of light. Capturing these shots well takes a lot of practice and skill, but it's a lot of fun to try.

Equipment:
- **Clear glass jar**
- **Camera**

Steps:

1. Start with fireflies in a jar. When you've got a captive audience, you know they won't escape, and you know where to focus your camera.

2. Use the highest ISO setting (film speed) you've got and a wide aperture. A wide aperture helps you capture all of the light from the flashes and the high ISO setting helps the camera register the light. Shutter speed is less[1] important, so try experimenting with different longer durations.

3. A cable shutter release[2] or a shutter remote is useful. That way you don't have to click the shutter button (and you won't inadvertently vibrate the camera). If you can set your camera up to take photos automatically, that's even better.

4. If your first results aren't what you hoped, keep trying. Capturing photographs of fireflies isn't easy, but when you're successful, it's very rewarding.

Project 4
Make Your Flashlight a Firefly Call

Just as birds respond to whistled calls, fireflies will sometimes respond to a flashlight that mimics their flashing pattern. Doing so is pretty simple—simply repeat the flashing pattern you see—but here are a few tips to make it more likely you'll get a return "call."

Steps:

1. Don't aim your flashlight[1] at a firefly. That could disorient the fireflies or cause them to fly away. Instead, direct your flashlight straight into the air.

2. Use an LED flashlight[2] covered with blue or red cellophane. These colors won't ruin your night vision; LED flashlights are also reported to be more effective than flashlights with traditional bulbs.

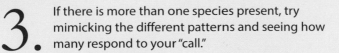

Equipment:
- Flashlight or LED flashlight
- Red or blue cellophane
- Paper and pencil for taking notes of the different flashing patterns

3. If there is more than one species present, try mimicking the different patterns and seeing how many respond to your "call."

4. Keep track of how many different species are present; such data can often be submitted to Citizen Science Projects (see page 86).

Project 5
Temperature Variance

Firefly flashing patterns vary by temperature. Flash rates drop along with the temperature, so a firefly will flash more when it is 70 degrees outside than if it is 55. If you are careful, you can even test this at home.

Note: Be very careful when carrying out this experiment. Only make the temperatures a few degrees warmer or colder than room temperature. Sudden temperature differences can harm the fireflies.

(ex. 72°)

Jar 1
room temperature

Steps:

1. Capture 3 fireflies and put 1 into each jar, then measure your room temperature and how often the firefly in Jar #1 flashes.

2. Place Jar #2 in water that is 1 or 2 degrees cooler than room temperature.

Equipment[1]:

- **Three fireflies of the same species**
- **Three jars**
- **Two large pots, preferably a double boiler or something that can hold a good amount of water**
- **A thermometer**
- **A stopwatch**
- **A kitchen sink**

(ex. 70°)

(ex. 74°)

Jar 2
cooler

Jar 3
warmer

3. Place Jar #3 in water that is 1 or 2 degrees warmer than room temperature.

4. Hold the jars in the water a bit, then time the flashing patterns. Can you see a difference?

Project 6

Firefly Tourism!

If you're really interested in fireflies, consider taking a road trip to see the synchronous fireflies in Great Smoky[1] Mountains National Park. The dates vary somewhat each year (as fireflies are not exactly predictable), but the fireflies are usually visible sometime between late May and late June; check with the park for specific dates for that year.

Great Smoky Mountains National Park
107 Park Headquarters Road
Gatlinburg, TN 37738
Visitor Information Recorded Message (865) 436-1200
Main website: www.nps.gov/grsm/index.htm
Firefly-specific website: www.nps.gov/grsm/naturescience/fireflies.htm

TIP: The park offers shuttles from the Sugarland Visitor Center to the firefly viewing areas. However, due to the popularity of the firefly viewing, you need to obtain a parking pass for the Sugarland Visitor Center in advance to be allowed on the shuttle. (Walk-ins are not allowed, in other words.) Parking passes are available at www.recreation.gov, but they go quickly (usually within the day), so be sure to get your passes that day. Note: registered campers at the park are also allowed on the shuttles.

Project 7

Help Stop Light Pollution

Much of firefly habitat—including most of the eastern U.S.—is affected by severe light pollution. In the areas with the worst light pollution, most of the stars in the night sky are lost in the glare. This doesn't only affect would-be stargazers and astronomers; it also harms fireflies. It's not hard to see why, given the firefly's unique light-based mating strategy.

Thankfully, you can help make your yard[1] and community a better place for fireflies by removing unnecessary outdoor lighting and switching to dark-sky friendly lighting fixtures, such as downward-facing lights. Doing so is relatively cheap and easy—you often save on your power bill—and you'll make your yard that much more welcoming to fireflies (and other nighttime creatures). You can also consider joining a dark-sky advocacy group—such as the International Dark Sky Association (www.darksky.org), which strives to create "dark-sky certified" communities parks and locales.

How bad is light pollution? During the blackout on the East Coast in 2003, 911 operators received dozens of frantic calls about an "unknown light in the sky." Some even claimed it was a UFO. It wasn't. They'd lived with light pollution for so long that they simply didn't recognize the light of the Milky Way.

Project 8
Citizen Science

When it comes to fireflies, some of the basic details (their range, which species live where) are in question. After all, fireflies are difficult to study, as they simply don't last that long. What's more, they are found across a wide range of habitats and there are only so many firefly researchers (and so much funding to go around). That's one reason Citizen Science projects, such as the Firefly Watch, are important. Created as a collaboration between Tufts University, the Museum of Science (Boston) and Fitchburg State University, the project enables volunteers to act as stand-in scientists, enabling a much wider collection of data.

Similar citizen-driven projects in other branches of science have been quite successful. Examples include Galaxy Zoo (www.galaxy zoo.org) and Planet Hunters (www.planethunters.org) in astronomy, the Great Backyard Bird Count (www.birdsource.org/gbbc) in ornithology, and bio-blitzes (www.bellmuseum.umn.edu/ ResearchandTeaching/BioBlitz/) in ecology and biology.

So if you're interested in fireflies, don't hesitate to sign up today. You can do so here: https://legacy.mos.org/fireflywatch/sign_up

Citizen Science projects have led to some impressive discoveries. The Galaxy Zoo project has helped catalog over 100,000 objects, leading to a number of important papers. The Planet Hunters project has even discovered the existence of extrasolar planets (planets that are not found in our solar system).

Recommended Reading—Books

Eisner, Thomas. Secret Weapons: *Defenses of Insects, Spiders, Scorpions, and Other Many-Legged Creatures.* (2007).

Harvey, E. Newton. *Bioluminescence.* (1952).

Harvey, E. Newton. *A History of Luminescence from the Earliest Times until 1900.* (1957). Free ebook available here: http://archive.org/details/historyoflumines00harv

Harvey, E. Newton. *The Nature of Animal Light.* (1920). Free ebook available here: www.gutenberg.org/files/34450/34450-h/34450-h.htm

Wilson, Thérèse and J. Woodland Hastings. *Bioluminescence: Living Lights, Lights for Living.* (2012).

Recommended Reading—Selected Papers

Arnett, Ross H., Jr., Michael C. Thomas, Paul E. Skelley, and J. Howard Frank. *American Beetles, Volume II: Polyphaga: Scarabaeoidea through Curculionoidea.* Boca Raton: CRC Press, 2010. http://books.google.com/books?id=gmgA0uxYhL0C&dq=Ellychnia corrusca glows as larvae&source=gbs_navlinks_s (accessed July 31, 2013)

Bitler, Barbara, and W. D. McElroy. "The preparation and properties of crystalline firefly luciferin." *Archives of Biochemistry and Biophysics.* no. 2 (1958): 358-368. www.sciencedirect.com/science/article/pii/0003986157902126 (accessed July 30, 2013)

Coblentz, William. *A Physical Study of the Firefly.* Washington, DC: Carnegie Institution of Washington, 1912. http://books.google.com/books?id=12wUAAAAYAAJ&pg=PP1

Daston, M. M., and J. B. Copeland. "The luminescent organ and sexual maturity in *Dyakia striata." Institute of Malacology.* no. 1 (1999): 9-19. http://cat.inist.fr/?aModele=afficheN&cpsidt=4824311 (accessed July 31, 2013)

Drees, B. M., and John Jackman. *Field Guide to Texas Insects.* Houston: Gulf Publishing Company, 1999. https://insects.tamu.edu/fieldguide/cimg290.html (accessed July 31, 2013)

Eisner, Thomas, et al. "Firefly 'femmes fatales' acquire defensive steroids (lucibufagins) from their firefly prey." *Proceedings of the National Academy of Science.* no. 18 (1997): 9723-9728. www.pnas.org/content/94/18/9723 (accessed July 31, 2013)

Eisner, Thomas, et al. "Two Cases of Firefly Toxicosis in Lizards." *Journal of Chemical Ecology*. no. 9 (1999): 1981-1986.
http://diyhpl.us/~bryan/papers2/paperbot/Firefly Toxicosis in Lizards.pdf (accessed July 31, 2013)

Frick-Ruppert, Jennifer, and Joshua J. Rosen. "Morphology and Behavior of *Phausis reticulate* (Blue Ghost Firefly)." *Journal of the NC Academy of Science*. no. 4 (2008): 139-147.
http://dc.lib.unc.edu/cdm/ref/collection/jncas/id/3883 (accessed July 31, 2013)

Ghiradella, Helen, and John Schmidt. "Fireflies at One Hundred Plus: A New Look at Flash Control." Integr. Comp. Biol. no. 3 (2004): 203-212.
http://icb.oxfordjournals.org/content/44/3/203.full.pdf html (accessed August 1, 2013)

Haddock, Steven, Mark Moline, and James Case. "*Bioluminescence in the Sea*." Annual Review of Marine Science. (2009): 443-493.
http://adsabs.harvard.edu/abs/2010ARMS....2..443H (accessed July 31, 2013)

Jusoh, Wan, et al. "Distribution and Abundance of *Pteroptyx* Fireflies in Rembau-Linggi Estuary, Peninsular Malaysia." *Environment Asia*. (2010): 56-60.
www.academia.edu/924360/Distribution_and_Abundance_of_Pteroptyx_Fireflies_in_Rembau-Linggi_Estuary_Peninsular_Malaysia (accessed July 31, 2013)

Pedigo, Larry, and Marlin Rice, *Entomology and Pest Management (6th Edition)*. Upper Saddle River: Prentice Hall, 2008.

Lee, John. Department of Biochemistry and Molecular Biology, University of Georgia, "A History of Bioluminescence." (accessed July 31, 2013)

Lewis, Sara, Lynn Faust, and Raphaël De Cock. "The Dark Side of the Light Show: Predators of Fireflies in the Great Smoky Mountains." *Psyche*. (2012).
http://ase.tufts.edu/biology/labs/lewis/publications/documents/2012LewisPsyche.pdf (accessed July 31, 2013)

Lloyd, J. E. "Studies on the flash communication system in *Photinus* fireflies." University of Michigan Museum of Zoology, Miscellaneous Publications, 130. (1966): 1-93.
http://deepblue.lib.umich.edu/handle/2027.42/56374 (accessed August 2, 2013)

Luk, S. P. L., Marshall, S. A., and Branham, M. A. 2011. "The Fireflies (Coleoptera; Lampyridae) of Ontario." *Canadian Journal of Arthropod Identification*. no. 16, 2 June 2011. www.biology.ualberta.ca/bsc/ejournal/lmb_16/lmb_16.html, doi: 10.3752/cjai.2011.16

Michael, T.P., Breton, G., Hazen, S.P., Priest, H., Mockler, T.C., et al. "A Morning-Specific Phytohormone Gene Expression Program underlying Rhythmic Plant Growth." 2008. *PLoS Biol* 6(9): e225. doi:10.1371/journal.pbio.0060225

Moiseff, Andrew, and Jonathan Copeland. "Firefly Synchrony: A Behavioral Strategy to Minimize Visual Clutter." *Science*. no. 5988 (2010): 181. www.sciencemag.org/content/329/5988/181.abstract?sid=4e04cf76-b362-4195-b1f0-7219d84bcea8 (accessed July 31, 2013)

Rooney, Jennifer, and Sara Lewis. "Notes on the Life History and Mating Behavior of *Ellychnia Corrusca* (Coleoptera: Lampyridae)." *Florida Entomologist*. no. 3 (200): 324-334. http://journals.fcla.edu/flaent/article/view/59556 (accessed July 30, 2013)

Seliger, H. H., J. B. Buck, W. G. Fastie, and W. D. McElroy. "The Spectral Distribution of Firefly Light." *Journal of General Physiology*. no. 1 (1964): 95-104. www.ncbi.nlm.nih.gov/pmc/articles/PMC2195396/ (accessed July 31, 2013)

Sivinksi, John. "The Nature and Possible Function of Luminescence in Coleoptera Larvae." *The Coleopterist's Bulletin*. no. 2 (1981).

Trimmer, Barry A., et al. "Nitric Oxide and the Control of Firefly Flashing." *Science*. no. 5526 (2001): 2486-2488. *Science* 29 June 2001. (accessed July 31, 2013)

Recommended Reading—Websites

Bioluminescent Beetles
www.bioluminescentbeetles.com

The Coleopterist's Bulletin (available for free)
www.jstor.org/action/showPublication?journalCode=colebull

The Entomological Association of America
www.entsoc.org/resources/faq/

Firefly.org
www.firefly.org

Firefly Photos by Terry Priest
www.frfly.com

Firefly Watch—Featured Research
Museum of Science, Boston
https://legacy.mos.org/fireflywatch/featured_research

"The Firefly Files." Branham, Marc. The Ohio State University.
http://hymfiles.biosci.ohio-state.edu/projects/FFiles

Science (many older journal articles are free) www.sciencemag.org

Companies that Produce Bioluminescence Kits

Carolina Scientific
www.carolina.com

Empco
http://empco.org/

Sea Farms
http://seafarms.com

Citizen Science Groups

BOINC
http://boinc.berkeley.edu

Firefly Watch, Museum of Science (Boston)
https://legacy.mos.org/fireflywatch/

Galaxy Zoo
www.galaxyzoo.org

International Dark Sky Association
www.darksky.org

Planet Hunters
www.planethunters.org

Scistarter
www.scistarter.com

Zooniverse
www.zooniverse.org/

Glossary

aggressive mimicry: when a predator mimics or masquerades as something else in order to deceive prey. The *Photuris* "Femme Fatale" fireflies are an example. They mimic *Photinus* males in order to lure them to their death

aposematism: when an organism warns off would-be predators. This usually indicates that the organism is distasteful or potentially harmful and would not be worth pursuing for the predator. This often occurs via warning coloration—bright colors or patterns that make their presence obvious and warn others to stay away. The firefly's light signals are an example of aposematism

bioluminescent organism: lifeforms that produce their own light via internal chemistry

binomial nomenclature: a system that uses a two-part name to identify an organism. Such names are unique and universal—no other organism shares it and the name is used worldwide. Often referred to as a scientific name, these names consist of a genus and a species. *Photinus pyralis*—the Common Eastern Firefly—is an example

Blue Ghost Firefly: the common name for *Phausis reticulata,* a tiny continuously glowing firefly that often glows in shades of blue or green

bugs: an informal name for all insects, even spiders, which aren't insects at all. Technically speaking, bugs refer to a specific order of insects—Hemiptera—which includes aphids and cicadas, among many other insects

chemical: a substance with a fixed composition and properties. Chemicals can't be separated by physical methods. Water (H_2O) is a good example: If you melt an ice cube, then freeze it again, you haven't changed the fundamental composition. To do so, you need a chemical reaction. Electrolysis is a common example; by running an electrical current through water, you can split water (H_2O) into oxygen and hydrogen

chemistry: the physical science that studies chemical elements and chemical compounds, and their reactions and properties

coleoptera a taxonomic order of insects that includes all beetles, including fireflies

endothermic a chemical reaction that absorbs energy

entomologist a scientist who studies insects

exothermic a chemical reaction that emits energy

enzyme large molecules that play many roles in living organisms; luciferase is the enzyme that helps fireflies light up

flies: flies are members of the insect order Diptera, which includes such familiar insects as houseflies, mosquitoes and gnats; fireflies aren't actually flies at all (they are beetles)

glowworms: a common—but misleading!—name for fireflies, which are insects, not worms

Great Smoky Mountains National Park: a national park in Tennessee and North Carolina that is known for its stunning displays of synchronous fireflies

habitat loss: when human development alters or destroys an animal's natural habitat; habitat loss is likely a problem for some firefly species

hotaru: the Japanese word for firefly; fireflies are very important symbols in Japan, and *hotaru* festivals are held each year

insect: a class of invertebrates that have three main body segments (head, thorax and abdomen), six legs, one or two pairs of wings, and a pair of antennae

insecticides: products that kill insects; some broad-spectrum insecticides kill many different types of insects, including fireflies

ISO: a setting on a digital camera that corresponds to film speed; faster film speed is more receptive to light

lantern organs: the light-producing organs located on the underside of a firefly

larva: the juvenile form of an insect

light pollution: when excess light from towns and cities lights up the night sky; this "glare" can affect fireflies and can even make it harder to see the stars

luciferase: an enzyme that helps catalyze (speed up) the light-producing reaction in fireflies

luciferin: the chemical compound that helps fireflies light up

peroxisomes: the storage tanks that hold the chemicals that help fireflies light up

pheromones: chemicals that insects (and other organisms) use to communicate; some firefly species don't light up to attract a mate; they use pheromones instead

Photinus: a genus of fireflies that includes the Common Eastern Firefly—the most common firefly in North America

photocytes: the reaction chambers of the firefly's lantern organs

Photuris: a genus of fireflies often referred to as the "Femme Fatales" because they mimic the signals of other fireflies in order to attract—and then eat—them

Pteroptyx: a genus of fireflies found in Southeast Asia; they are famous for blinking synchronously (all at once)

pupa: the third phase of an firefly's life; in the pupa phase, the firefly transforms from a larva into an adult

Pyractomena: a genus of fireflies found in North America that are known for their amber-colored flash

reflex bleeding: a firefly defense mechanism where the firefly begins bleeding when threatened; as fireflies are toxic, the bad-tasting blood seems to scare away would-be predators

synchronous: occurring at the same time

taxonomy: a system used for classification; in biology, the taxonomic system helps scientists categorize life

thanatosis: when an animal plays dead; named for the Greek god of the dead, Thanatos

Bibliography

The Basics

[1] Centre for Ecology and Hydrology (CEH) Wallingford, "Taxonomy of Bioluminescent Beetles" www.bioluminescentbeetles.com/taxonomy/ (accessed July 30, 2013)

Insects Are Everywhere

[1] Entomological Society of America, "Frequently Asked Questions on Entomology" www.entsoc.org/resources/faq/ (accessed July 30, 2013)

[2] Mertens, Randy. "New insect species identified on campus" *Mizzou Weekly*, May 2, 2013. http://mizzouweekly.missouri.edu/archive/2013/34-29/new-bug/index.php (accessed July 30, 2013)

Fireflies That Don't Glow

[1] Rooney, Jennifer, and Sara Lewis. "Notes on the Life History and Mating Behavior of *Ellychnia Corrusca* (Coleoptera: Lampyridae)" *Florida Entomologist*. no. 3 (200): 324-334. http://journals.fcla.edu/flaent/article/view/59556 (accessed July 30, 2013)

[2] Bush-Pirkle, Merrik. "Spanish Fly Beetles Use Sex and Subterfuge to Infiltrate Bee's Nests" Published by the Public Affairs Office at San Francisco State University, Diag Center. Last modified May 05, 2000. www.sfsu.edu/news/prsrelea/fy99/101.htm (accessed August 1, 2013)

What's In a Name?

[1] Brown, R.W. *Composition of scientific words: A manual of methods and a lexicon of materials for the practice of logotechnics.* Washington, DC: Smithsonian Institution Press, 1956. Quoted in BugGuide.net: http://bugguide.net/node/view/12418 (accessed July 30, 2013)

[2] Ibid.

[3] Weller, Susan. "Minnesota Profile Fireflies (family Lampyridae)" *Minnesota Conservation Volunteer*, July 2002. www.dnr.state.mn.us/volunteer/julaug02/fireflies.html (accessed July 30, 2013)

Four Phases of Life

[1] Drees, B.M., and John Jackman. *Field Guide to Texas Insects*. Houston: Gulf Publishing Company, 1999. https://insects.tamu.edu/fieldguide/cimg290.html (accessed July 31, 2013)

Firefly Eggs and Larvae

[1] McKenzie, J. 2001. "*Photinus pyralis*" (Online), Animal Diversity Web. http://animaldiversity.ummz.umich.edu/accounts/Photinus_pyralis/ (accessed July 30, 2013)

[2] Oba, Yuichi, M. Furuhashi, M. Bessho, S. Sagawa, H. Ikeya, and S. Inouye. "Bioluminescence of a firefly pupa: involvement of a luciferase isotype in the dim glow of pupae and eggs in the Japanese firefly, *Luciola lateralis*" *Photochemical & Photobiological Sciences*. no. 5 (2013): 854-863. doi: 10.1039/c3pp25363e (accessed August 8, 2013)

Larvae Lunch

[1] Lyon, William. Ohio State University Extension Fact Sheet, "Firefly HYG–2125-95," http://ohioline.osu.edu/hyg-fact/2000/2125.html (accessed July 30, 2013)

Why Larvae Glow

[1] Sivinksi, John. "The Nature and Possible Function of Luminescence in Coleoptera Larvae" *The Coleopterist's Bulletin*. no. 2 (1981).

[2] Ibid.

Home Sweet Home

[1] Branham, Marc. The Ohio State University, "The Firefly Files." http://hymfiles.biosci.ohio-state.edu/projects/FFiles/frfact.html (accessed July 30, 2013)

[2] Ibid.

Lantern Organs

[1] Ghiradella, Helen, and John Schmidt. "Fireflies at One Hundred Plus: A New Look at Flash Control" Integr. Comp. Biol. no. 3 (2004): 203-212. http://icb.oxfordjournals.org/content/44/3/203.full.pdf html (accessed August 1, 2013)

Luciferin: A Chemical Named for the Morning Star

[1] Maas, Anthony. *The Catholic Encyclopedia*, Vol. 9. New York: Robert Appleton Company, 1910. www.newadvent.org/cathen/09410a.htm (accessed July 30, 2013)

[2] Bitler, Barbara, and W.D. McElroy. "The preparation and properties of crystalline firefly luciferin" *Archives of Biochemistry and Biophysics*. no. 2 (1958): 358-368. www.sciencedirect.com/science/article/pii/0003986157902126 (accessed July 30, 2013)

The Science of Fireflies

[1] Lee, John. Department of Biochemistry and Molecular Biology, University of Georgia, "A History of Bioluminescence" (accessed July 31, 2013)

How Do Fireflies Control When They Light Up?

[1] Trimmer, Barry A., et al. "Nitric Oxide and the Control of Firefly Flashing" *Science*. no. .5526 (2001): 2486-2488, 29 June 2001. (accessed July 31, 2013)

[2] Ghiradella, Helen, and John Schmidt. "Fireflies at One Hundred Plus: A New Look at Flash Control" *Integr. Comp. Biol.* no. 3 (2004): 203-212. http://icb.oxfordjournals.org/content/44/3/203.full.pdf html (accessed August 1, 2013)

How Bright are Fireflies?

[1] Coblentz, William. *A Physical Study of the Firefly*. Washington, DC: Carnegie Institution of Washington, 1912. http://books.google.com/books?id=12wUAAAAYAAJ&pg=PP1

[2] Levy, Hazel. *University of Florida Book of Insect Records* (Chapter 29). Gainesville: University of Florida, Gainesville, 1998.

Why Do Fireflies Look So Bright?

[1] Burton, Maurice, and Robert Burton. *International Wildlife Encyclopedia*. New York City: Marshall Cavendish Corporation, 2002. http://books.google.com/books?id=0gsPc5lk7_UC&lpg=PA827&ots=mv77CGkZyj&dq=pyrophorus noctilucus candlepower&pg=PA827

The Science Behind the Firefly Color Palette

[1] Seliger, H.H., J.B. Buck, W.G. Fastie, and W.D. McElroy. "The Spectral Distribution of Firefly Light" *Journal of General Physiology*. no. 1 (1964): 95-104. www.ncbi.nlm.nih.gov/pmc/articles/PMC2195396/ (accessed July 31, 2013)

Mating: Why They Glow

[1] Branham, Marc. "How and why do fireflies light up?" *Scientific American*, September 5, 2005. www.scientificamerican.com/article.cfm?id=how-and-why-do-fireflies (accessed July 31, 2013)

A World of Variation

[1] Sharp, K. 2001. "*Photuris versicolor*" (Online), Animal Diversity Web. http://animaldiversity.ummz.umich.edu/accounts/Photuris_versicolor/ (accessed July 31, 2013)

[2] Lloyd, J.E. "Studies on the flash communication system in *Photinus* fireflies" University of Michigan Museum of Zoology, Miscellaneous Publications, 130. (1966): 1-93. http://deepblue.lib.umich.edu/handle/2027.42/56374 (accessed August 2, 2013)

[3] Branham, Marc. "How and why do fireflies light up?" *Scientific American*, September 5, 2005. www.scientificamerican.com/article.cfm?id=how-and-why-do-fireflies (accessed July 31, 2013)

Other Creatures with Bioluminescence

[1] Haddock, Steven, Mark Moline, and James Case. "Bioluminescence in the Sea" *Annual Review of Marine Science*. (2009): 443-493. http://adsabs.harvard.edu/abs/2010ARMS....2..443H (accessed July 31, 2013)

[2] Daston, M.M., and J.B. Copeland. "The luminescent organ and sexual maturity in *Dyakia striata*" *Institute of Malacology*. no. 1 (1999): 9-19. http://cat.inist.fr/?aModele=afficheN&cpsidt=4824311 (accessed July 31, 2013)

Lunch—and Something Extra

[1] Eisner, Thomas, et al. "Firefly 'femmes fatales' acquire defensive steroids (lucibufagins) from their firefly prey" *Proceedings of the National Academy of Science*. no. 18 (1997): 9723-9728. www.pnas.org/content/94/18/9723 (accessed July 31, 2013)

[2] Eisner, Thomas. Jacob Gould Schurman Professor Emeritus of Chemical Ecology, "Thomas Eisner: List of Publications." www.nbb.cornell.edu/neurobio/eisner/eisnerbiblio.pdf (accessed August 1, 2013)

Predator and Prey
[1] Lewis, Sara, Lynn Faust, and Raphaël De Cock. "The Dark Side of the Light Show: Predators of Fireflies in the Great Smoky Mountains" *Psyche*. (2012). http://ase.tufts.edu/biology/labs/lewis/publications/documents/2012LewisPsyche.pdf (accessed July 31, 2013)

Reflex Bleeding: A Strange Survival Strategy
[1] Lewis, Sara, Lynn Faust, and Raphaël De Cock. "The Dark Side of the Light Show: Predators of Fireflies in the Great Smoky Mountains" *Psyche*. (2012). http://ase.tufts.edu/biology/labs/lewis/publications/documents/2012LewisPsyche.pdf (accessed July 31, 2013)

[2] Ibid.

Don't Feed Fireflies to Your Pet … or Your Brother!
[1] Eisner, Thomas, et al. "Two Cases of Firefly Toxicosis in Lizards" *Journal of Chemical Ecology*. no. 9 (1999): 1981-1986. http://diyhpl.us/~bryan/papers2/paperbot/Firefly Toxicosis in Lizards.pdf (accessed July 31, 2013)

A Cold Light
[1] Branham, Marc. "How and why do fireflies light up?" *Scientific American*, September 5, 2005. www.scientificamerican.com/article.cfm?id=how-and-why-do-fireflies (accessed July 31, 2013)

The Femme Fatales
[1] Museum of Science (Boston), "Firefly Watch: Types of Fireflies." https://legacy.mos.org/firefly-watch/types_of_fireflies (accessed July 31, 2013)

The Femme Fatales Find a Mate
[1] Srour, Marc. Teaching Biology, "Fireflies (Coleoptera: Lampyridae)." http://bioteaching.wordpress.com/2012/01/06/fireflies-coleoptera-lampyridae/ (accessed July 31, 2013)

The *Pyractomena* Fireflies
[1] Firefly.org, "Types of Fireflies." www.firefly.org/types-of-fireflies.html (accessed July 31, 2013)

[2] Arnett, Ross H. Jr., Michael C. Thomas, Paul E. Skelley, and J. Howard Frank. *American Beetles, Volume II: Polyphaga: Scarabaeoidea through Curculionoidea*. Boca Raton: CRC Press, 2010. http://books.google.com/books?id=gmgA0uxYhL0C&dq=Ellychnia corrusca glows as larvae&source=gbs_navlinks_s (accessed July 31, 2013)

A Synchronized Show in the Great Smokies
[1] Great Smoky Mountains National Park, "Synchronous Fireflies." www.nps.gov/grsm/nature-science/fireflies.htm (accessed July 31, 2013)

Why Synchronize?
[1] Moiseff, Andrew, and Jonathan Copeland. "Firefly Synchrony: A Behavioral Strategy to Minimize Visual Clutter" *Science*. no. 5988 (2010): 181. www.sciencemag.org/content/329/5988/181.abstract?sid=4e04cf76-b362-4195-b1f0-7219d84bcea8 (accessed July 31, 2013)

[2] Yamasaki, Alisa. "Firefly festivals in the summer capture spirit of growing up in old Japan" *The Japan Times*. June 14, 2013. www.japantimes.co.jp/culture/2013/06/14/events/firefly-festivals-in-the-summer-capture-spirit-of-growing-up-in-old-japan/

Other Synchronous Firefly Varieties

[1] Jusoh, Wan, et al. "Distribution and Abundance of *Pteroptyx* Fireflies in Rembau-Linggi Estuary, Peninsular Malaysia" *Environment Asia.* (2010): 56-60. www.academia.edu/924360/Distribution_and_Abundance_of_Pteroptyx_Fireflies_in_Rembau-Linggi_Estuary_Peninsular_Malaysia (accessed July 31, 2013)

The Blue Ghost Firefly: A Steady Light in the Dark

[1] Frick-Ruppert, Jennifer, and Joshua J. Rosen. "Morphology and Behavior of Phausis reticulate (Blue Ghost Firefly)" *Journal of the NC Academy of Science.* no. 4 (2008): 139-147. http://dc.lib.unc.edu/cdm/ref/collection/jncas/id/3883 (accessed July 31, 2013)

[2] Ibid.

Concerns for the Present (and the Future)

[1] Mydans, Seth. "Firefly populations are disappearing" *The New York Times*, November 20, 2008. www.nytimes.com/2008/11/20/world/asia/20iht-fireflies.1.17990392.html?_r=1& (accessed July 31, 2013)

Problems with Insecticides

[1] Clemson University, "Vanishing Firefly Project: Environmental Impacts." www.clemson.edu/public/rec/baruch/firefly_project/firefly_impacts.html (accessed July 31, 2013)

[2] Hopwood, Jennifer, et al. The Xerxes Society for Invertebrate Conservation, "Are Neonicotinoids Killing Bees?" Last modified 2012. http://ento.psu.edu/publications/are-neonicotinoids-killing-bees (accessed July 31, 2013)

[3] Blessing, Arlene, ed. "Pesticides and Pest Prevention Strategies for the Home, Lawn, and Garden, PPP-34" Purdue Pesticides Programs, Purdue University Cooperative Extension Service. www.ppp.purdue.edu/Pubs/ppp34.html (accessed July 31, 2013)

A Lot We Still Don't Know

[1] Kay, S., T.P. Michael, G. Breton, S.P. Hazen, H. Priest, T.C. Mockler, et al. 2008. A Morning-Specific Phytohormone Gene Expression Program underlying Rhythmic Plant Growth. *PLoS Biol* 6(9): e225. doi:10.1371/journal.pbio.0060225

Mythology

[1] American Museum of Natural History, "Creatures of Light: Firefly," www.amnh.org/exhibitions/past-exhibitions/creatures-of-light/creatures/firefly (accessed July 31, 2013)

[2] Turpin, Tom. Purdue Extension Service, "On Six Legs: Myths About Insects Thick as Fleas on Dogs." Last modified May 09, 2002. www.agriculture.purdue.edu/agcomm/newscolumns/archives/OSL/2002/May/020509OSL.html (accessed July 31, 2013)

Fireflies, Poets and Playwrights

[1] Craig, W.J., ed. "[Pericles, Prince of Tyre, Act II, Scene II]" *The Complete Works of William Shakespeare*. London: Oxford University Press: 1914; Bartleby.com, 2000. www.bartleby.com/70

Firefly Distribution

[1] Luk, S.P.L., S.A. Marshall, and M.A. Branham. 2011. "The Fireflies (Coleoptera; Lampyridae) of Ontario." *Canadian Journal of Arthropod Identification*. no. 16, 2 June 2011. Available online at www.biology.ualberta.ca/bsc/ejournal/lmb_16/lmb_16.html, doi: 10.3752/cjai.2011.16

[2] Firefly Watch Project, Museum of Science, Boston. https://legacy.mos.org/fireflywatch/images/MOS_FFW_2013_Sightings.pdf

Firefly Season Calendar

[1] Based on data from the Firefly Watch Project, Museum of Science, Boston. https://legacy.mos.org/fireflywatch/images/MOS_FFW_2013_Sightings.pdf

Finding the Right Habitat

[1] Drees, B.M., and John Jackman. *Field Guide to Texas Insects*. Houston: Gulf Publishing Company, 1999. https://insects.tamu.edu/fieldguide/bimg153.html (accessed July 31, 2013)

[2] McKenzie, J. 2001. "*Photinus pyralis*" (Online), Animal Diversity Web. http://animaldiversity.ummz.umich.edu/accounts/Photinus_pyralis/ (accessed July 31, 2013)

Firefly Flashing Patterns

[1] Chart based on data from the Firefly Watch Project, Museum of Science, Boston. https://legacy.mos.org/fireflywatch/flash_charthttps://legacy.mos.org/fireflywatch/flash_chart

Project: A Home Away From Home

[1] Firefly.org, "How to Catch Fireflies." www.firefly.org/how-to-catch-fireflies.html (accessed July 31, 2013)

Project: Firefly Photography

[1] Adams, Kevin. "Photographing Fireflies in a Jar" Digital After Dark (blog), August 19, 2010. www.kadamsphoto.com/nightphotography/photographing-fireflies-in-a-jar/ (accessed July 31, 2013)

[2] Ibid.

Project: Make Your Flashlight a Firefly Call

[1] Firefly.org, "How to Catch Fireflies." www.firefly.org/ (accessed July 31, 2013) how-to-catch-fireflies.html

[2] Ibid.

Project: Temperature Variance

[1] Museum of Science (Boston), "Firefly Watch: Flash Chart." https://legacy.mos.org/fireflywatch/flash_charthttps://legacy.mos.org/fireflywatch/flash_chart (accessed July 31, 2013)

Project: Firefly Tourism!

[1] Great Smoky Mountains National Park, "Synchronous Fireflies." www.nps.gov/grsm/nature-science/fireflies.htm (accessed July 31, 2013)

Project: Help Stop Light Pollution!

[1] Walthall, Vanessa. University of Florida IFAS Extension Service, "How to Have Fireflies in Your Backyard." Last modified April 09, 2009. http://leon.ifas.ufl.edu/News_Columns/2009/040909g.pdf (accessed July 31, 2013)

Photo Credits

14, (left inset) Pennsylvania Department of Conservation and Natural Resources - Forestry Archive, Bugwood.org, (right inset) Joseph Berger, Bugwood.org. **15**, (inset) Pennsylvania Department of Conservation and Natural Resources - Forestry Archive, Bugwood.org. **16**, (all) Professor Stephen Luk **18**, (pupa) Professor Stephen Luk **19**,(right) Gerald J. Lenhard, Louisiana State University, Bugwood.org, (left) Courtesy of Professor Xinhua Fu. **22**, Courtesy of Professor Stephen Luk **23**, (main) Susan Ellis, Bugwood.org. **26**, Courtesy of Professor Helen Ghiradella and her colleagues. **44**, Professor Stephen Luk **46**, (inset) Professor Stephen Luk **47**, (inset) Professor Stephen Luk **49**, (inset) Professor Stephen Luk **53**, (both) Courtesy of Professor Jennifer E. Frick-Ruppert and her colleagues. **56**, Courtesy of Professor Steve Kay and his colleagues. **102**, Kayli Schaaf.

35, Based on data from *Studies on the Flash Communication System in Photinus Fireflies*, James E. Lloyd, University of Michigan Museum of Zoology, 1966.

64–65, Map based on data from the Firefly Watch Project, Museum of Science, Boston. Accessed July 31, 2013. https://legacy.mos.org/fireflywatch

66–67, Map based on data from the Firefly Watch Project, Museum of Science, Boston. Accessed July 31, 2013. https://legacy.mos.org/fireflywatch

70, Charts based on data from the Firefly Watch Project, Museum of Science, Boston. Accessed July 31, 2013. https://legacy.mos.org/fireflywatch/flash_chart

The (unaltered) images on the following pages are licensed according to the GNU Free Documentation License version 1.2, which is accessible here: www.gnu.org/licenses/old-licenses/fdl-1.2.txt

20, "Raupe" vs Schnecke" by Wikipedia user Heinz Albers, www.heinzalbers.org; available at: http://commons.wikimedia.org/wiki/File:Raupe_schnecke.JPG

36, (right) "NZ Glowworm" by Wikipedia User Markrosenrosen; available at: http://commons.wikimedia.org/wiki/File:Nz_glowworm.jpeg

The (unaltered) images on the following pages are licensed according to the Creative Commons Sharealike 2.0 License, which is accessible here: http://creativecommons.org/licenses/by-sa/2.0/

21, (inset) "Leuchtkäfer - Firefly" by Wikipedia User NEUROtiker; available at: http://commons.wikimedia.org/wiki/File:Leuchtkäfer_-_Firefly.JPG

84, "Numerous fireflies glowed above the stream" by Flickr User T. Kiya; available at: www.flickr.com/photos/cq-biker/5832210597

The (unaltered) images on the following pages are licensed according to the Creative Commons Attribution 2.0 License, which is accessible here: http://creativecommons.org/licenses/by-sa/2.0/us/

27, (inset) "Firefly on the Screen Door" by Flickr User Slgckgc; available at: www.flickr.com/photos/slgc/7411319414

About the Author

Brett Ortler is an editor at Adventure Publications. While at Adventure, he has edited dozens of books, including many field guides and nature-themed books. His own work appears widely, including in *Salon*, *The Good Men Project*, *The Nervous Breakdown*, *Living Ready* and in a number of other venues in print and online. He lives in the Twin Cities with his wife and their young children.